Pollution Control

London: H M S O

Researched and written by Reference Services, Central Office of Information.

© Crown copyright 1993
Applications for reproduction should be made to HMSO.
First published 1993

ISBN 0 11 701777 9

HMSO publications are available from:

HMSO Publications Centre
(Mail, fax and telephone orders only)
PO Box 276, London SW8 5DT
Telephone orders 071-873 9090
General enquiries 071-873 0011
(queuing system in operation for both numbers)
Fax orders 071-873 8200

HMSO Bookshops
49 High Holborn, London WC1V 6HB   071-873 0011
Fax 071-873 8200 (counter service only)
258 Broad Street, Birmingham B1 2HE   021-643 3740   Fax 021-643 6510
Southey House, 33 Wine Street, Bristol BS1 2BQ
0272 264306   Fax 0272 294515
9-21 Princess Street, Manchester M60 8AS   061-834 7201   Fax 061-833 0634
16 Arthur Street, Belfast BT1 4GD   0232 238451   Fax 0232 235401
71 Lothian Road, Edinburgh EH3 9AZ   031-228 4181   Fax 031-229 2734

HMSO's Accredited Agents
(see Yellow Pages)

and through good booksellers

Contents

Acknowledgments

The Central Office of Information would like to acknowledge the assistance of the following in the preparation of this book: the Department of the Environment; the Department of Trade and Industry; the Department of Transport; the Ministry of Agriculture, Fisheries and Food; the Northern Ireland Office; the Scottish Office; the Welsh Office; the Overseas Development Administration and the National Rivers Authority.

Introduction

For more than a century Britain[1] has been developing policies to protect the environment against pollution from industry and other sources. Laws were introduced at an early stage to control air and water pollution, and to conserve wildlife, the landscape, historic monuments and buildings. These have been revised regularly to meet changing circumstances and newly-perceived threats. Administrative and enforcement arrangements have also been updated to ensure that these policies are implemented effectively.

Increasingly, the control of pollution, and the protection of the environment more generally, is undertaken on an international basis. Britain has entered into many treaties and conventions and is bound by other supranational undertakings such as European Community directives. These international initiatives tackle a wide range of problems, such as ozone depletion, marine pollution and the threat of climate change.

This book describes the policies and legislation by which Britain seeks to control pollution.

---

[1]The term 'Britain' is used informally to mean the United Kingdom of Great Britain and Northern Ireland. 'Great Britain' comprises England, Scotland and Wales.

Policy Framework

Government policies and legislation to control pollution are developed and implemented across a number of departments and agencies. Since 1990 a series of White Papers have been published at annual intervals, bringing together a range of policies for the protection of the environment.

This Common Inheritance

The White Paper *This Common Inheritance* (see **Further Reading, p. 75**), published in September 1990, was the first comprehensive statement by the Government of its policy on issues affecting the environment. It summarised more than 350 proposals for tackling such diverse issues as global warming, pollution control, the regulation of land use and planning, the rural economy, the countryside and wildlife. There were also pledges of action on heritage, air quality and pollution, noise, water, hazardous substances, waste and recycling, and separate sections on specific initiatives in Scotland, Wales and Northern Ireland. Important initiatives specifically in the field of pollution control included:

—more widely available information on the results of monitoring of air pollutants;

—a new expert panel to advise on air quality standards;

—an increase in the maximum fine that magistrates' courts could impose for water pollution offences;

—the introduction of a system of recycling credits (see p. 60);

—strengthened penalties for noise nuisance; and

—government pressure for European Community (EC) action on energy efficiency and the adoption of a system of integrated pollution prevention and control.

On climate change, the White Paper reaffirmed Britain's willingness to return carbon dioxide (CO_2) emissions to 1990 levels by the year 2005,[2] provided other countries took similar action. Many of the actions that the Government has since taken to improve the environment were announced in *This Common Inheritance*.

In September 1991 a first anniversary progress report was published, which reviewed progress and set out commitments to more than 400 further actions to protect the environment and integrate environmental considerations into the work of government as a whole. These included:

—the appointment of an environmental expert on all relevant public bodies and groups;

—strategies in every government department for environmental management of buildings and land, to be introduced by the end of 1992;

—regular reporting on environmental performance;

—the development of proposals to phase out all remaining uses of polychlorinated biphenyls and the publication of guidance for their safe disposal;

—the introduction of car exhaust emission MOT testing by November 1991 (see p. 36); and

—the creation of a unified Environment Agency (see p. 14).

[2]This target date has subsequently been brought forward to the year 2000—see p. 27.

A further update appeared in October 1992. This started by looking at the outcome of the 'Earth Summit' (see p. 6) and the steps that Britain is taking to implement the obligations it entered into. Overall, the report covered action on more than 440 environmental commitments made since the original White Paper was published. It also made 441 pledges of further action, including:

—new air quality standards, to be introduced in 1993 for ground-level ozone, benzene and carbon monoxide and by 1994 for sulphur dioxide and 1,3 butadiene;

—the extension of the air quality monitoring network to all major cities by 1997;

—work towards achieving 2,000 megawatts of extra capacity from combined heat and power schemes by the year 2000;

—statutory water quality objectives to be introduced from 1993;

—the publication of government guidance on nature conservation and the planning process by the end of 1992; and

—tighter new standards for waste sites from April 1993.

Economic Instruments
The major policy sections of the 1992 White Paper contained a description of economic instruments that could be used to gain environmental benefits. Among the possible economic instruments looked at were:

—a carbon or energy tax, which is being considered by the European Commission as a means of encouraging energy efficiency and the use of low- or non-carbon fuels and technologies, hence reducing carbon dioxide emissions (see p. 29);

—higher fuel prices and the introduction of road pricing to discourage urban congestion and ensure that costs incurred by road users reflect the full cost of their journeys, including environmental costs;

—new charging regimes in fields such as water abstraction so as to encourage the efficient use of resources; and

—a range of instruments to encourage recycling of solid waste, including waste collection charges, product levies and waste disposal charges.

Environmental Information

Britain is committed to the principle that the public should have rights of access to environmental information. Legislation has established registers of environmental information to which the public have free access, covering such matters as control of radioactive substances; industrial processes which could cause significant pollution to air, land and water; litter control; waste management licences; and water pollution and supply. New regulations came into force in December 1992, giving the public additional rights to information not held on statutory registers. These regulations apply to a wide range of public bodies with environmental responsibilities, including government departments and agencies, local authorities and organisations with environmental responsibilities, such as the Countryside Commission. Safeguards are provided for the protection of confidential information, together with a power to make reasonable charges for supplying information.

A new statistical report, *The UK Environment* (see **Further Reading**, p. 75), giving comprehensive data on the British environment, has recently been published by the Department of the Environment. It has been designed so as to be easy to understand. Most aspects of the environment are covered, including climate; air quality and factors affecting the global atmosphere; soil and land use; water resources and quality; the marine environment; coastal erosion; wildlife; waste; and radioactivity. The report also covers the links between environment and health, the effects of mankind's activities on the environment, public attitudes to environmental issues and expenditure on the environment. The report supplements the Department's *Digest of Environmental Protection and Water Statistics,* which will continue to be published annually. The new report will be updated at regular intervals, probably every two or three years.

International Action

Britain participated fully in the United Nations Conference on Environment and Development (UNCED), popularly known as the Earth Summit, held in Rio de Janeiro in June 1992. Among the agreements reached were a framework convention on climate change (see p. 27), a convention on biological diversity, 'Agenda 21' (an action framework for the twenty-first century), a declaration setting out clear principles for sustainable development, and a statement of principles for the management of forests. In Rio de Janeiro Britain also launched:

—a Darwin Initiative for the Survival of Species;[3]

—the Technology Partnership Initiative, to ensure that developing countries could share the benefits of technology through partnership with British companies; and

—Partnerships for Change '93, an international conference due to be held in Manchester in September 1993, to enable non-governmental organisations to exchange experience of sustainable development.

[3]For further information on this and other features of Britain's conservation policies, see *Conservation* (Aspects of Britain: HMSO,1993).

Increasingly, much of Britain's legislation on the environment, including the control of pollution, is being developed in collaboration with other member states of the EC and organisations such as the Organisation for Economic Co-operation and Development and the United Nations and its agencies. The EC has adopted over 300 pieces of environmental legislation applicable in Britain, covering a wide range of problems, including air and water pollution, waste management, toxic and dangerous substances, nature conservation and the environmental aspects of development projects.

During the British Presidency of the EC in the second half of 1992, measures were adopted on transfrontier shipments of waste, ozone-depleting substances and exhaust emissions from light commercial vehicles. Progress was made on a new voluntary eco-management and audit scheme, and on measures on the wildlife trade and on packaging and packaging waste. The Community also addressed the follow-up to the UNCED meeting, especially on climate change issues, and approved its Fifth Action Programme, mapping out the Community's environmental strategy for the rest of the 1990s. The British Presidency also pressed for further action to improve both the integration of environmental concerns into other policies and the implementation and enforcement of Community legislation. Britain is also taking a leading position in the development of a European eco-labelling scheme (see p. 17).

Assistance to Developing Countries and Eastern Europe

Cleaning up the environment is not cheap—for example, it is estimated that control of pollution costs Britain between 1 and 1.5 per cent of its gross national product. In its domestic environmental policies, Britain supports the principle that the polluter should pay, and is examining various means to promote this approach. It is recognised, however, that developing countries do not have the same ability to pay the financial costs of dealing with environmental problems, since their economies are not as strong as those of developed countries. As well as drawing up policies for preventing pollution at home, therefore, Britain supports the efforts of countries overseas to protect their environment. This assistance takes many forms—for example, support for programmes designed to conserve rhino and elephants in Africa—but it also includes assistance specifically on the control of pollution.

Through the Technical Co-operation Training Programme of the Overseas Development Administration (ODA), Britain helps trainees from developing countries to acquire skills in areas such as environmental science, soil conservation and water resource management. Britain has a budget separate from the rest of the aid programme specifically designed to address global environmental issues. It provides technical advice on environmental policy and supports the transfer of environmentally-benign technology to developing countries. This help can include financial assistance—for example, in replacing gases which damage the ozone layer (see p. 31). Britain has offered over £40 million for the Global Environment Facility, jointly operated by the World Bank, the United Nations Environment Programme and the United Nations Development Programme, during its 1991–1994 programme. In 1992 Britain gave £4.5 million, £500,000 more than in 1991, to the Environment Fund, run by the United Nations Environment Programme to help developing countries meet the cost of new environmental challenges. This made Britain the second biggest contributor to the Fund.

The newly-democratic countries of eastern Europe have in many cases severe environmental problems that have to be tackled. To assist in this, in June 1991 Britain announced the setting up of an Environmental Know How Fund. This will spend £5 million over three years in providing training and the transfer of knowledge and expertise with a specifically environmental flavour. Its work began in April 1992: missions to identify suitable projects in Bulgaria, the former Czechoslovakia, Hungary, Poland and Romania took place in 1992. Further missions to the Commonwealth of Independent States and the Baltic states are also being planned.

Precautionary Action

There are several emerging environmental concerns where it is not yet certain how serious the problems are and further scientific investigation is needed.

Britain believes, however, that in such cases it is not right to delay taking action until proof of the scale of the problem emerges. Where there are significant risks of damage to the environment, the Government is prepared to take precautionary action to limit the use of potentially dangerous materials or stop the spread of potentially harmful pollutants, if the balance of likely costs and benefits justifies it.

Administrative and Legislative Framework

Executive responsibility for pollution control is divided between local authorities and central government agencies. Central government makes policy, exercises general budgetary control, promotes legislation and advises pollution control authorities on policy implementation. The most recent major piece of environmental legislation is the Environmental Protection Act 1990, which introduced more comprehensive standards of pollution control and new arrangements to promote conservation. The Water Act 1989—now consolidated into the Water Resources Act 1991—also strengthened the framework for water pollution control.

Government Departments

The Secretary of State for the Environment has general responsibility for co-ordinating the work of the Government on environmental protection. The Department of the Environment therefore carries out a very wide range of functions in the control of pollution. In Scotland, Wales and Northern Ireland the respective Secretaries of State are responsible for pollution control co-ordination within their countries. They also carry out many of the roles that fall to other departments in England.

Other government departments also play an important part in the control of pollution. The Ministry of Agriculture, Fisheries and Food has a major role in the prevention of pollution from farming activities. It also has important responsibilities in the control and registration of pesticides, the licensing of sea disposal of waste and the protection of the food chain from radioactive materials or other hazardous substances. The Department of Transport maintains a special Marine Pollution Control Unit (see p. 49) that is responsible for detecting and deterring illegal marine pollution, and also assists in clear-up operations in the event of an incident. From April 1992, the Department of Trade and Industry took over responsibility for Britain's energy industry (apart from the Energy Efficiency Office, which became part of the Department of the Environment) when the former Department of Energy was wound up. The Department of Trade and Industry also collaborates with the Department of the Environment in supporting the research effort of British industry (see pp. 19–20).

Local Authorities

In addition, local authorities and a wide range of voluntary organisations are involved in environmental protection. Local authorities have important duties and powers. They are responsible for matters such as:

—collection, disposal and regulation of domestic wastes (see pp. 54–7);

—keeping the streets free from litter (see p. 57);

—control of air pollution from domestic and from many industrial premises (see p. 24); and

—the abatement of general nuisances, and of noise from most sources.

These responsibilities are divided between the different tiers of local government. For example, in England, waste collection is undertaken by district councils and waste disposal by county councils. However, there are some exceptions; in London and other English metropolitan areas, both of these functions are generally discharged by the London boroughs and metropolitan districts, although some single-purpose waste disposal authorities have been set up in these areas, while councils often pool their resources to make joint arrangements for refuse disposal. In Wales, district councils are responsible for both collection and disposal of waste. In Scotland, the district and islands councils act as waste collection, disposal and regulation authorities. The islands councils also act as river purification authorities for their areas. In Northern Ireland, refuse collection and disposal are the responsibility of district councils.

The present structure of local government in Britain is under review. A move towards the more widespread establishment of single-tier authorities is being considered in England by the Local Government Commission, while the Government proposes to introduce such unitary authorities in Scotland and Wales. Such changes would affect the present distribution of local government responsibilities. The proposed establishment of an Environment Agency (see p. 14) would also result in local authorities giving up their present responsibility for waste regulation.

National Rivers Authority

The National Rivers Authority (NRA) was set up following the privatisation of the water supply and sewerage industry. Previously, state-owned water authorities were responsible both for the supply of water to consumers and the regulation and enforcement of water quality. While the former function remains with the water supply companies, the NRA was established to handle the latter function. It is responsible for the control of water pollution and the quality of rivers, lakes, estuaries and coastal waters in

England and Wales (see pp. 40–2); it also has duties relating to the management of water resources; salmonid, eel and freshwater fisheries; shellfisheries; and flood protection. The NRA expects to spend £123 million on grant-aided services in 1992–93. Flood defence work is paid for by precepts on local authorities, while a proportion of pollution control work is funded by charges.

The regulation of the quality of drinking water supplied to consumers, however, remains the responsibility of the Drinking Water Inspectorate, which is part of the Department of the Environment. Local authorities have responsibilities in relation to the monitoring of private water supplies and are also empowered to monitor the quality of drinking water supplied through the mains.

Her Majesty's Inspectorate of Pollution

In England and Wales Her Majesty's Inspectorate of Pollution (HMIP) has an important role in the control of releases to land, air and water from certain industrial processes through the mechanism of integrated pollution control (IPC—see p. 15). HMIP was formed in 1987 by the merger of three existing inspectorates—the Industrial Air Pollution Inspectorate (formerly part of the Health and Safety Executive), the Hazardous Waste Inspectorate and the Radiochemical Inspectorate—together with a new Water Pollution Inspectorate. Its main function is to implement the IPC provisions of the Environmental Protection Act 1990 through the authorisation and monitoring of prescribed processes. It has powers to prosecute operators of processes which breach release limits set in their authorisations. As HMIP's responsibilities have been widened in recent years with the introduction of IPC, its staff complement has also been increased considerably to ensure that it can discharge its functions effectively.

HMIP is increasingly working in co-operation with parallel enforcement agencies in other EC member states. To promote links between these bodies, a permanent Network of Enforcement Agencies has been established. Its first meeting, hosted by HMIP during Britain's EC Presidency, was held in November 1992 in Chester. It was attended by more than 60 representatives of environmental enforcement agencies from all 12 member states. The meeting agreed to:

— exchange information and experience through regular meetings and via a network of contact points in each country;

— examine means of exchanging staff between agencies;

— establish working groups to examine specific matters of joint interest; and

— provide advice on the enforcement aspects in the development of new environmental legislation.

Britain has also advocated the establishment of an Audit Inspectorate, a small body which would scrutinise the ability of member states' enforcement agencies to meet their obligations under EC legislation. This would not usurp the role of national inspectorates, but would help ensure that laws drawn up at Community level were applied in an effective and consistent manner.

The Government is committed to introducing early legislation to establish a single Environment Agency for England and Wales. The new agency would bring together all the present functions of HMIP and the NRA, together with the waste regulation functions of local authorities.

Scotland and Northern Ireland

In Scotland, the river purification authorities have statutory responsibility for water pollution control. These consist of seven river purification boards, which cover the mainland, and the three islands councils, each of which acts as the river purification authority for its area. The seven mainland boards forecast expenditure of £11.8 million in 1992–93. Her Majesty's Industrial Pollution Inspectorate (HMIPI) undertakes functions in Scotland similar to those of HMIP in England and Wales. Like HMIP, HMIPI is also working increasingly in co-operation with enforcement agencies in other EC countries. It was a founder member of the Network of Enforcement Agencies. A new Scottish environment agency is being planned to bring together the responsibilities of HMIPI and the river purification authorities and some of the environmental protection functions of local authorities.

In Northern Ireland, water quality is monitored by the Environment Service of the Department of the Environment for Northern Ireland.

Integrated Pollution Control

Under the Environmental Protection Act 1990, a system of IPC is being phased in to control releases to air, water and land from certain categories of industries. The most potentially harmful processes are 'scheduled' for IPC, and require authorisation from HMIP. In granting authorisation for releases under IPC, HMIP requires the use of the best available techniques not entailing excessive cost to prevent or minimise polluting releases and to ensure that any releases are made harmless. Authorisations under IPC and applications for authorisation—which contain comprehensive details about the processes operated—are placed in a public register, although the applicant can ask HMIP that certain information is excluded from the register on grounds of commercial confidentiality. Applications must be advertised and members of the public have a right to make representations to HMIP.

In Scotland, HMIPI administers IPC jointly with the river purification authorities. In Northern Ireland broadly similar controls are exercised by the Environment Service, which is part of the Department of the Environment for Northern Ireland, and proposals are being formulated for the introduction of a new system of air pollution control. The Government is pressing for the introduction of IPC on the British model within the EC.

In November 1992 the Government put forward proposals for consultation on the gradual introduction in England and Wales of an annual inventory of releases of polluting substances from industrial processes. This would provide aggregated data on releases from industrial processes subject to IPC, as well as from premises regulated under the Radioactive Substances Act 1960. Information would be available at both national and local levels to ensure that it was of the greatest value to interested parties. The introduction of an inventory in this form would not place any additional burden on industry, as the information contained is already required by HMIP to monitor compliance with authorisations under IPC.

Other Bodies

An Advisory Committee on Business and the Environment (ACBE) was set up in May 1991 to assist the Government on environmental issues relevant to business. Its terms of reference are to:

—provide for a dialogue between Government and business on environmental matters;

—help mobilise the business community, in liaison with other appropriate organisations, in demonstrating good environmental practice and management; and

—provide a link with international business initiatives on the environment.

The ACBE's activities include working on two booklets published in November 1992—*A Practical Energy-saving Guide for Smaller Businesses* and *A Guide to Environmental Best Practice in Company Transport*—in collaboration with government departments.

The British Standards Institution, which sets quality standards for a very wide range of products, has published a new standard environmental management system (BS 7750), modelled along the lines of the widely-adopted quality standard BS 5750. Like this latter standard, BS 7750 is designed to be applicable to all types of organisation, and also to be compatible with proposed EC regulations on eco-management and auditing. The central concept in BS 7750 is environmental auditing; this aims to be a management tool to help safeguard the environment by facilitating management control of environmental practices and assessing compliance with company policies, including the meeting of regulatory requirements.

Britain has taken a leading role in helping to develop a European Community eco-labelling scheme. The purpose of the scheme is to help consumers identify products which do least harm to the environment, and thereby encourage the development and manufacture of such products. A new body—the UK Ecolabelling Board—was set up in July 1992 to run the scheme in Britain. It is hoped that the first labels will start to appear on goods in the shops towards the middle of 1993.

There are also a wide range of pressure groups in Britain that are actively concerned with the protection of the environment and the prevention of pollution.

Environmental Research

Government research into environmental protection is generally co-ordinated by the Department of the Environment. The Government stressed in *This Common Inheritance* that such work is essential to its environmental policies in order to ensure that decisions are based on the best scientific evidence. It has been estimated that total government spending on environmental research and development in 1991–92 was £300–400 million, including work in areas such as renewable energy. The Department of the Environment expects to spend about £87 million in 1992–93 on research into subjects including:

—climate change;

—atmospheric pollution and its monitoring;

—toxic chemicals and genetically modified organisms (GMOs);

—waste disposal; and

—water quality and health.

Other departments have substantial programmes. The Ministry of Agriculture, Fisheries and Food spends over £50 million a year on research and development to draw up policies for the protection of the agricultural environment. The Scottish Office Agriculture and Fisheries Department and other official bodies such as the NRA and the Meteorological Office also have important programmes.

An independent standing Royal Commission on Environmental Pollution advises the Government on national and international matters concerning the pollution of the environment, on the adequacy of research and on the future possibilities of danger to the environment. So far it has produced 16 reports.

Environmental Technology Innovation Scheme

One of the research initiatives launched under the White Paper was the Environmental Technology Innovation Scheme (ETIS), which combined existing schemes to promote innovation and competitiveness in the environmental technology industry. The scheme is jointly sponsored by the Department of the Environment and the Department of Trade and Industry. Grant assistance of up to 50 per cent of a project's costs can be made for industrial research that is not yet competitive. Generally the scheme aims to support collaborative ventures, but support of up to 25 per cent of a project's costs may be available from the Department of the Environment in certain technology areas for a company undertaking work alone.

Among the projects that have benefited from help under the three-year ETIS programme since it was launched in October 1990 are:

—development work on a thermal oxidiser to generate power from exhaust air contaminated with solvents from the printing and coating industries—thus reducing emissions of such solvents— supported by a grant of £268,000;

—a programme to develop new techniques to treat the gases arising from waste incineration, supported by a grant of £264,000;

—research into the recycling of waste from the production of mineral wool, which aims to reduce emissions of carbon dioxide and the demand for landfill space, supported by a £43,000 grant; and

—a scheme, which is receiving a £42,000 grant, to assist in the control of harmful emissions from brick, ceramics and glass-making by developing a continuous analyser to measure fluoride concentrations.

Environmental Management Options Scheme

The Department of Trade and Industry also supports an Environmental Management Options Scheme (DEMOS). Like ETIS, the scheme was launched in October 1990 and will run until October 1993. It aims to help industry adopt environmental best practices in three main areas: cleaner technology, recycling and treatment of wastes and effluents.

An example of a project helped under the DEMOS initiative is a £1.3 million scheme, announced in September 1992, to turn domestic waste and sewage sludge into a useful compost through aerobic fermentation. This project, which will receive £409,000 of government grant over three years, is a collaboration between South West Water Services, Devon County Council and Focsa UK Ltd, a multinational company involved in handling waste. If successful, it will help provide an alternative to the sea dumping of sewage sludge, which Britain will cease in 1998 (see p. 45), and will also relieve the problem posed to the disposal of municipal waste by landfill shortages.

Research Councils

Basic and strategic research is carried out by five government-funded research councils. All have a role in environmental protection research, but particularly important is the Natural Environment Research Council (NERC), which has a science budget allocation of £130 million in 1992–93, plus expected receipts of about £43 million from commissioned research and other income. The NERC undertakes and supports research in the environmental sciences and funds postgraduate training. Its programmes encompass the marine, earth, terrestrial, freshwater, polar and atmospheric sciences. The NERC stresses international collaborative work on global environmental issues alongside the Department of the Environment's research programmes. For example, it is helping to develop global atmospheric climate models and is strengthening atmospheric research in the Arctic. A major new research programme, the Terrestrial Initiative in Global Environmental Research (TIGER), has been started, so that the impact of climate change on Britain and elsewhere can be foreseen. The NERC also co-ordinates the development and operation of the Environmental Change Network, which is funded by the Department of the Environment, the Ministry of Agriculture, Fisheries and Food, and others.

The Agricultural and Food Research Council (AFRC) spends about £25 million a year on research aimed at understanding the interaction between agriculture and the environment. The Ministry of Agriculture, Fisheries and Food commissions a large programme of research into environmental protection from AFRC institutes and Horticulture Research International. The AFRC is a co-sponsor, with the NERC, of a co-ordinated programme, costing £1.2 million over three years, on pollutant transport in soils and rocks. With the Science and Engineering Research Council, it sponsors the Clean Technology programme, which seeks to develop new technologies to prevent or reduce pollution from agriculture. The AFRC has also launched a major co-ordinated programme on biological adaptation to global environment change, costing £8 million over four years. This looks at biological responses to altered climatic variables, including carbon dioxide, water, temperature and ultraviolet radiation exposure. It complements, and is linked with, the NERC's TIGER programme.

Air Pollution

Air pollution can cause problems over a wide area—from local difficulties caused by photochemical smog affecting a particular city, to global problems such as climate change and ozone depletion caused by emissions of certain gases. Britain is therefore party to a number of international agreements on air pollution, such as the Montreal Protocol (see p. 31) and the United Nations Economic Commission on Europe's Convention on Long Range Transboundary Air Pollution.

Clean Air Acts

Responsibility for clean air rests primarily with local authorities. Considerable progress has been made towards the achievement of cleaner air and a better environment in the past 30 years or so.

Smoke from coal fires used to pose a considerable problem in Britain's cities. In extreme cases, such as the notorious London smog of 1952, human lives could be threatened. Since the passage of the Clean Air Acts 1956 and 1968, however, there have been great improvements. Average winter visibility in London, for example, has increased from about 2.5 km to about 6 km (1½ miles to 4 miles). It no longer has the dense smoke-laden 'smogs' of the 1950s and, in central London, winter sunshine has increased by about 70 per cent since 1958. Similar improvements have been achieved in other cities.

Most of Britain's smoke control areas, within which smokeless fuels can be burned but other coal can only be burned in appliances capable of burning it smokelessly, were declared soon after the legislation was brought in. Some new smoke control areas, however, are still being established; the area affected by such orders increased by 30 per cent between 1979 and 1989. Within newly-declared smoke control areas, local authorities are able to give householders grants towards the installation of new boilers and fires. Controls are being introduced to stop the sale of unauthorised fuels within existing smoke control areas.

Table 1: Estimated Black Smoke Emissions 1970–90

	Thousand tonnes		*Thousand tonnes*
1970	1,028	1981	530
1971	921	1982	529
1972	802	1983	511
1973	807	1984	474
1974	771	1985	545
1975	672	1986	574
1976	647	1987	528
1977	665	1988	522
1978	627	1989	502
1979	641	1990	453
1980	560		

Source: Department of the Environment

Straw and Stubble Burning

In the past, straw and stubble burning after the harvest has caused a considerable nuisance in agricultural areas of Britain. It was therefore decided in 1989, after a high level of public complaint, to ban this practice in England and Wales. Farmers were given three years in which to find alternative methods of disposing of their unwanted straw. In the three harvests up to 1992, straw burning fell by 60 per cent as farmers responded to the new situation. The 1993 harvest will see the introduction of a ban on the burning of crop residues, with some minor exceptions. These include the residues from minor crops such as hops, lavender and reeds, where no public nuisance has been caused by their burning, and the residues from linseed which, because of the nature of the plant, are particularly difficult to dispose of except by burning. The exemption for linseed will be reviewed after the 1995 harvest.

Industrial Processes

Those industrial processes with the greatest potential for harmful atmospheric emissions are controlled under the Environmental Protection Act 1990, in England and Wales by HMIP; they are becoming subject to IPC. Processes with a lesser potential for air pollution require approval from local authorities under the local authority air pollution control (LAAPC) regime also introduced by the 1990 Act. A far higher number of premises are subject to LAAPC than are controlled under IPC.

Under the Clean Air Acts, emissions of dark smoke from any trade or industrial premises or from the chimney of any building are prohibited, and new furnaces must be capable as far as is practicable of smokeless operation. The 1990 Act also gave local authorities in England and Wales streamlined powers to deal with statutory nuisances, including smoke, dust and smells. It is an offence under the Health and Safety at Work etc Act 1974 for a company to fail to use the best practicable means to prevent the emission of noxious substances into the atmosphere; fines for this can be up to £20,000 if imposed by a magistrate's court, or unlimited if imposed by the Crown Court. HMIP has been successful in bringing prosecutions for breaches of this provision, with substantial fines sometimes being imposed. The first prosecution under the Environmental Protection Act was successfully brought by HMIP in March 1993.

Monitoring of Air Quality

The Government monitors air quality in Britain at over 400 sites. The most important pollutants are automatically checked for, including smoke, sulphur dioxide (SO_2), lead, nitrogen dioxide (NO_2), acid rain and carbon monoxide (CO). Air pollution data from the Department of the Environment's monitoring network are released to the national press and television each day as Air Quality Bulletins. These give the concentrations of three main pollutants—ozone, NO_2 and SO_2—and grade air quality on a scale between 'very poor' and 'very good'.

The data from the network of air quality monitoring sites are made available to members of the public on a special telephone number, in several newspapers and on videotext systems. From May 1992, the two existing helplines were replaced by a single free service. When air quality is forecast to be bad, warnings can be given to the public in the weather forecast bulletins. In October 1992 the Government accepted the recommendation of its Advisory Group on the Medical Aspects of Air Pollution Episodes that warnings should be issued when concentrations of sulphur dioxide in the air are expected to exceed 400 parts per thousand million, a level that could cause asthma sufferers to experience discomfort.

Local authorities also monitor air quality. For example, Cardiff City Council makes information on air quality within Cardiff available on a daily basis.

A comprehensive government review of urban air quality was announced in January 1992. Three independent committees of experts will advise on aspects of the problem, and will set guidelines and targets for air quality in Britain. The automatic air quality monitoring network is being upgraded at a cost of £10 million.

Climate Change

The earth is warmed naturally by the atmosphere trapping the heat of the sun. Indeed, without this 'greenhouse effect', there could be no life on earth. The earth's climate has changed considerably in the past, for example during the Ice Ages in prehistoric times, as well as less dramatically in recorded history. But recently scientists have become concerned that man-made atmospheric emissions of 'greenhouse gases', such as carbon dioxide, methane and nitrous oxide, are leading to greater concentrations of these gases in the atmosphere. This in turn could lead to a greater propensity for the earth's atmosphere to retain heat, with considerable potential effects on the world's climate.

In 1988 the United Nations Environment Programme and the World Meteorological Organisation established the Intergovernmental Panel on Climate Change to consider climate change and possible responses to it. Three working groups were established; Britain chairs the group which assesses the scientific evidence on climate change. Other working groups were to examine the economic and social impact of climate change, and possible policy options to combat it.

The British-led working group concluded, in a report published in May 1990, that man-made emissions would lead to additional warming of the earth, and that, without any change in emissions, global average temperature would increase by 0.3° C a decade. This would be faster than at any time over the past 10,000 years. Such changes could have major effects on the world, for example causing sea levels to rise by about 6 cm (2½ inches) a decade due to the thermal expansion of the oceans and melting of the polar ice caps. It could also have important effects on agriculture, soils, pests, diseases and wildlife.

Framework Convention

In December 1990 the United Nations set up an Intergovernmental Negotiating Committee to draw up a framework convention on climate change in time for the UNCED meeting in Rio de Janeiro in June 1992. Britain played a leading role in the negotiations towards the convention, which was duly opened for signature at the UNCED meeting. The convention commits all signatories to devising and reporting on the measures that they propose to take to combat climate change. In addition, it commits developed countries to action aimed at returning emissions of carbon dioxide and other greenhouse gases to 1990 levels by the year 2000. This commitment will be reviewed at the first meeting of all parties to the convention, which will probably be in 1995, and again in 1998. Britain sees the outcome of the negotiations as a significant first step in the global response to climate change; the review process incorporated into the convention is one of its most important aspects. Britain was one of over 150 countries that signed the convention at the UNCED meeting.

Together with the other members of the 'Group of Seven' leading industrial nations,[4] Britain is committed to ratifying the convention, and to publishing its first national programme to implement it by the end of 1993.

[4]These are: Britain, Canada, France, Germany, Italy, Japan and the United States.

Implementation

The implementation programme will build on measures that have already been taken to limit emissions; these are concentrated on the promotion of energy efficiency in power generation and use. An Energy Saving Trust was launched in May 1992, which brings together the Government, British Gas and the regional electricity companies to promote energy saving schemes. The Government has also launched a major advertising campaign to emphasise the environmental arguments for efficient use of energy in the home.

As part of the implementation programme, in December 1992 the Government published a discussion document aimed at encouraging a national debate on Britain's programme for limiting carbon dioxide emissions. It is based on figures published in October 1992, which show that carbon dioxide emissions in the year 2000 could be between 157 and 179 million tonnes, compared with 160 million tonnes in 1990. Under the study's central scenario, on which the Government's programme will be based, carbon dioxide emissions in 2000 would be 170 million tonnes, 10 million tonnes more than the 1990 figure.

Table 2: Carbon Dioxide Emissions by Sector, 1970–2000

	Final energy consumers, million tonnes of carbon			
	1970	1980	1990	2000*
Households	54	48	41	41
Industry	85	64	56	58
Commerce/public sector	20	23	24	27
Transport	23	29	38	45
Total	182	16	16	17

*projection   Source: Department of the Environment
Note: Differences between totals and the sums of their component parts are due to rounding.

The document identifies the further scope available for significant savings from energy efficiency measures. It recognises, however, that these are unlikely to achieve the full reductions needed, and that therefore the Government will have to introduce further measures, which could be regulatory or fiscal. Among the regulatory measures discussed are minimum efficiency standards for appliances and the fitting of speed limiters to cars. Fiscal possibilities include incentives to encourage energy-efficient investments, and measures to increase the price of energy, such as increases in road fuel duty or energy or carbon taxes.

In May 1992 the EC proposed a carbon or energy tax in order to limit carbon dioxide emissions in member states. Britain set up a working party to consider the details of the proposal during its presidency of the Community. This group reported progress on its continuing work to the council of finance ministers in December 1992. The proposal is one of the options considered in the discussion document, although the Government is not yet persuaded that the proposal should be adopted.

A number of other gases contribute to the threat of global warming, of which the most important are methane and nitrous oxide. Various measures are being taken to limit their emissions— for example, action is being taken to collect methane from many landfill waste sites, partly for safety reasons and partly to exploit the energy potential of the gas. The programme that Britain is required to draw up under the climate change convention will set out details of these measures.

International Assistance
Britain is also assisting developing countries in their response to the threat of climate change. The ODA funds various projects, for example:

—studies in Ghana and Kenya into the potential impact of climate change and into devising policies to tackle the problem;

—research on the economic effects of limiting carbon dioxide emissions in Zimbabwe; and

—a series of seminars on climate change for ministers and senior officials in developing countries.

Research

Britain is making a major research effort into global warming. The Hadley Centre for Climate Prediction and Research was opened in 1990 to build on the climate modelling programme of the Meteorological Office, at a cost of about £6 million a year to the Department of the Environment. The Department's total expenditure on research related to climate change will be over £14 million in 1993–94. Other bodies researching into the effects of climate change include the AFRC (see p. 21), which in June 1991 announced support of £4 million for new research on how plants and animals react to global environmental alterations such as changed temperatures. In addition, the Ministry of Agriculture, Fisheries and Food spent over £1 million in 1992–93 on studies into agriculture's contribution to greenhouse gas releases and the potential impact of climate change on crops, pests, diseases and soils.

Ozone Layer

The earth is protected from certain harmful wavelengths of ultra-violet radiation from the sun by a layer of ozone in the strato-sphere—between 10 km and 50 km (6 to 30 miles) above the surface of the earth. In recent years, concern has been growing that various gases, man-made or resulting from human activity, are harming the ozone layer. Such substances include chlorofluorocarbons (CFCs), which in the past have been widely used for applications such as insulation in refrigerators and air-conditioning plants and in the manufacture of expanded polystyrene. Observations have shown that a 'hole' appears seasonally in the ozone layer above the Antarctic every year; this was first observed in 1984 by scientists from the British Antarctic Survey. Continued thinning of the ozone layer could lead to health problems such as increased skin cancers. Britain is therefore committed to the earliest possible phasing out of all ozone-depleting substances, and has achieved some considerable success, for example reducing its consumption of CFCs by 50 per cent between 1985 and 1989. CFCs are also very powerful greenhouse gases, and so measures to reduce CFC usage will also help reduce the threat of climate change.

Montreal Protocol

Britain was one of the first 25 signatories to the Montreal Protocol, which deals with the protection of the ozone layer. It hosted the second meeting of the parties in London in June 1990, which substantially strengthened the protocol. Under the revised protocol, the supply of CFCs, halons and carbon tetrachloride was to be phased out by the year 2000 and 1,1,1 trichloroethane by 2005. There was provision for exemptions for any essential uses of halons. Stricter EC legislation was also introduced, requiring CFCs to be phased out within member states by mid-1997, with exemptions for any essential uses until the end of 1999. A financial mechanism was agreed to help developing countries to meet the requirements of the protocol; Britain has pledged a minimum of $9 million towards such assistance. The ODA has also encouraged developing countries to agree the protocol, for example by carrying out a global study on behalf of the United Nations Environment Programme into the costs of CFC substitution to developing countries, and by working with India in 1990–91 to examine strategies that it could adopt for phasing out ozone-depleting substances.

More recent scientific evidence nevertheless showed a need for further measures. The Montreal Protocol was reviewed again in November 1992, and Britain co-chaired the negotiations leading up to the review. The EC proposed that CFCs, halons, carbon tetrachloride and 1,1,1 trichloroethane should be phased out by the end of 1995. It also proposed that there should be tight controls on hydrochlorofluorocarbons (HCFCs—transitional substances with less ozone-depleting capacity than CFCs, which are needed in a number of areas if industry is to move away from CFCs quickly). Some other possible replacements for CFCs exist, but have potential problems. For example, hydrofluorocarbons do not damage the ozone layer but are powerful greenhouse gases, while substances such as ammonia and butane have problems with toxicity or flammability.

The review meeting agreed to revise the timetable for the phasing out of ozone-depleting substances, making it considerably quicker. A timetable for phasing out HCFCs was also agreed for the first time. Under the revised timetable:

—CFCs, carbon tetrachloride and 1,1,1 trichloroethane will be phased out by the end of 1995;

—halons will be phased out by the end of 1993;

—HCFCs will be phased out by 2030, with a limit on consumption levels from 1996 and substantial reductions from 2004 onwards; and

—a freeze at 1991 levels, from the end of 1994, on the production
of methyl bromide, which has only recently been recognised as
an ozone-depleting substance.

All these phase-out dates are subject to exemptions for essential
uses if the parties to the various agreements so decide. The EC has
subsequently agreed to phase out CFCs and carbon tetrachloride
by the end of 1994. Britain has offered $100,000 towards the
Montreal Protocol review process.

Public Awareness

The Government has tried to increase public awareness of the need
to protect the ozone layer. A free leaflet was launched in May 1992,
which tells consumers about products that still contain CFCs and
other substances harmful to the ozone layer. It also gives advice on
the proper disposal of old fridges and freezers that contain CFCs.

Emissions of Sulphur Dioxide and Oxides of Nitrogen

Sulphur dioxide and oxides of nitrogen (NO_x) are the main gases
that lead to acid rain. The main sources of these gases are power
stations and other large industrial plants which burn fossil fuels and
also, for oxides of nitrogen, motor vehicles. The damaging effect of
acid depositions from combustion processes on freshwaters and
soils has been demonstrated by scientific research. The
Government is spending about £10 million a year on an extensive
research programme into the causes and effects of acid rain, and the
likely results of possible abatement technologies.

Emissions of sulphur dioxide in Britain have been falling for
some years, from over 6 million tonnes in 1970 to under 3.7 million
tonnes in 1989, and the Government has initiated a substantial

programme to ensure that this fall continues. Emissions of oxides
of nitrogen from most sources have also been falling, although
increased emissions as a result of the growth in road transport have
meant that overall levels have increased somewhat in recent years.
Lower emissions of sulphur dioxide over the past 20 years have led
to the first signs of a decrease in acidification in some lochs in
south-west Scotland.

Table 3: UK Sulphur Dioxide Emissions 1970–90

	Million tonnes		*Million tonnes*
1970	6.424	1981	4.436
1971	6.057	1982	4.211
1972	5.785	1983	3.861
1973	6.005	1984	3.719
1974	5.511	1985	3.724
1975	5.370	1986	3.895
1976	5.184	1987	3.898
1977	5.159	1988	3.811
1978	5.225	1989	3.719
1979	5.541	1990	3.774
1980	4.898		

Source: Department of the Environment

Under the EC directive on the control of emissions from large
combustion plants such as power stations, the Government has
published a national plan setting out phased reductions in emissions from existing plants of oxides of nitrogen to 1998 and of sulphur dioxide to 2003. The reductions that the Government
proposes to make are higher than those required by the terms of the
directive, as shown in Table 4.

Table 4: Target Percentage Reductions from 1980 Baselines

	Sulphur dioxide		Oxides of nitrogen	
	EC requirement	*Target*	*EC requirement*	*Target*
1993	20	21	15	21
1998	40	45	30	35
2003	60	63	—	—

Source: Department of the Environment

Government figures, published in November 1992, show that Britain is meeting its planned reduction in sulphur dioxide and oxides of nitrogen and is on course to meet the targets for 1993. These figures have been sent to the European Commission on a plant-by-plant basis for the larger plants, as required by the directive. A variety of measures are being used by the electricity supply industry to meet their requirements in the plan, including:

— the installation of flue gas desulphurisation equipment at some power stations;

— installing special low-NO_x burners at power stations; and

— switching from coal to gas for power generation, as well as the greater use of low-sulphur coal.

Vehicle Emissions

Tighter controls on the amount of carbon monoxide (CO), hydrocarbons and oxides of nitrogen which can be emitted by petrol- and diesel-engined cars have been agreed in the European Community. These mean that, from the end of 1992, new petrol-engined cars have had to be fitted with catalytic converters. Emissions of gaseous pollutants from road vehicles generally increased between 1981 and 1990, although lead emissions have dropped greatly (see p. 37). The European Commission is considering a proposal for still stricter limits for cars for 1996 and 2000, for which a formal proposal is expected in 1993. Emissions from light-duty goods vehicles are being brought in line with those for passenger cars.

New, tighter standards have also been agreed by the EC for diesel-engined lorries and buses. These standards are being implemented in two phases:

—from October 1993, new vehicles will have to meet a substantially tighter limit on emissions of oxides of nitrogen;

—from October 1996 the limit on oxides of nitrogen will be reduced to about half its value prior to October 1993, and the limit on particulates will be similarly stringent to that to be applied in the United States from 1994, which is the tightest standard anywhere in the world.

Further limits for implementation in the year 2000 are to be discussed within the Community.

A check for emissions in the annual 'MOT' test of vehicle roadworthiness was introduced in November 1991 for petrol-engined cars, measuring carbon monoxide and hydrocarbons. In September 1992 the annual roadworthiness smoke check for heavy goods vehicles, buses and coaches was changed to an instrumented test, and the same type of test was implemented for diesel-engined passenger cars in January 1993. Similar arrangements are being introduced in Northern Ireland.

Lead

Atmospheric lead largely results from vehicle emissions. This is being combated through the promotion of unleaded petrol and reduced amounts of lead in leaded petrol. The amount of lead in the air in Britain has halved since the permitted lead content in petrol was reduced in 1986 from 0.4 to 0.15 grammes a litre; total emissions of lead from road vehicles in Great Britain fell by nearly 70 per cent between 1981 and 1990.

Britain took a leading role in negotiating an EC directive which required unleaded petrol to be available throughout the Community by October 1989. Almost all petrol stations in Britain now sell unleaded petrol, and new cars must be able to run on it. A substantial tax differential has been created in favour of unleaded fuel in recent years, so that it is significantly cheaper than leaded petrol. Demand for unleaded petrol has since risen rapidly; by March 1993 unleaded petrol accounted for half of all petrol deliveries in Britain. Unleaded petrol is also important in that cars fitted with catalytic converters must be run on unleaded fuel, as leaded petrol can damage the converters.

Volatile Organic Compounds
One of the major sources of volatile organic compounds (VOCs) is road traffic, although they also come from sources such as industry and the use of solvents. VOCs can pose a major pollution problem, especially by acting as a precursor of ground level ozone, which is formed by the action of sunlight on VOCs. Such photochemical smogs can harm human health, interfere with plant growth and damage building materials.

The Government has adopted a target of a 30 per cent reduction in emissions of VOCs by 1999, using 1988 levels as a baseline; this results from a United Nations Economic Commission for Europe protocol signed in November 1991. The protocol commits signatories, among other things, to:

—a 30 per cent reduction in VOC emissions between 1988 and 1999;

—apply national or international emission standards for new sources of VOCs;

—promote the use of products with a low or nil VOC content; and

—negotiate a second stage of VOC emission reductions, following more scientific work.

A consultation document on Britain's proposed strategy to meet this requirement was published in November 1992. This sets out the action already being taken or planned to meet Britain's obligations under the protocol. A government leaflet was launched in May 1992 to alert members of the public to the problem and to advise them about the most effective actions they could take to reduce photochemical smog.

Aviation

Internationally developed standards have been introduced to control the emission from civil aircraft of smoke, vented fuel and unburned hydrocarbons. Current indications are that aircraft contribute only a small amount to overall pollution, although the Government is funding further research and taking a leading role in European studies on this issue.

Water Pollution

As with air pollution, water pollution can have an international effect, for pollutants that enter inland waters may well eventually reach the sea. Britain applies strict controls on the emission of polluting matter to water, and is introducing the IPC system further to control more tightly the most harmful discharges. A 'Red List' of the most harmful substances—those most toxic, persistent and likely to accumulate in living organisms—has been drawn up, and inputs of these substances are subject to the strict controls of IPC. Considerable progress has been made in reducing inputs of such substances to the marine environment.

Freshwater Protection

In general, it is against the law to allow any polluting matter to enter water in Britain except in accordance with a legal authorisation or consent. In England and Wales the NRA is responsible for protecting water quality. Its principal method of controlling water pollution is through the regulation of all effluent discharges into inland and coastal waters (except those subject to IPC, which are controlled by HMIP). Discharge consents issued by the NRA specify what may be discharged and set limits on the volume and content of effluent, in order to achieve appropriate environmental quality standards. The NRA maintains public registers containing information about discharge consents and water quality.

In England and Wales the water industry is planning to spend some £14,000 million at 1989 prices up to the year 2000 on improving the sewer network and sewage treatment and disposal (see p. 45). Progressively higher treatment standards for industrial waste effluents, and measures being taken to combat pollution from agriculture, are expected to bring further improvements in water quality.

The Government is committed to meeting the requirements of a number of EC directives for the protection and improvement of water quality, for example, on the quality of surface water for abstraction for drinking water supply, the quality needed to support freshwater fisheries and shellfish, the quality of water for bathing areas (see p. 46) and the treatment of urban waste water.

Similar arrangements for the control of pollution apply in Scotland, where control is exercised by the river purification authorities. Sewage treatment and disposal come within the water and sewerage programme, which will total about £728 million in the three years to 1995–96. Unlike in England and Wales, the provision of water and sewerage services is a local authority function, although the Government is considering restructuring these services in the light of the proposed changes in local government.[5] The Environment Service is responsible for controlling water pollution in Northern Ireland.

Quality Objectives

The Government was empowered to introduce statutory water quality objectives (WQOs) for individual stretches of water under the Water Resources Act 1991, as well as to set up a new scheme for classifying water quality.

A government discussion paper on the implementation of statutory WQOs in England and Wales was published in December 1992. It proposed that statutory WQOs for rivers should be phased

[5]For information on the proposals on local government, see *Scotland* (Aspects of Britain: HMSO, 1993), p. 18.

A workman is dwarfed by a pipe ready for installation at the flue gas desulphurisation plant at the Drax B power station. When the plant is fully operational in 1996, it will remove about 90 per cent of the sulphur dioxide from the power station's emissions.

This award-winning device reclaims and recycles ozone-depleting CFC gases from old fridges and freezers that are being repaired. It enables engineers making home visits to retrieve the CFCs in the appliance–previously they might have been released into the atmosphere.

This machine has been developed to burn the gases–mostly methane– arising from rotting rubbish in landfill sites to generate electricity. It thus reduces emissions of a powerful 'greenhouse' gas while also conserving fossil fuel resources.

COI Pictures/B. Bell

The annual 'MOT' test now includes a check of exhaust emissions, the reliability of which depends upon accurate analysers being used at the 17,000 test centres. This specially-constructed testing chamber is used to calibrate those analysers, which has to be done four times a year.

COI Pictures/B. Bell

A scientist from the University of East Anglia checks a lake for man-made nutrients in effluents from agriculture and water treatment works. These nutrients can lead to rapid growth in algae, to the detriment of other aquatic life.

Recycling facilities now often allow members of the public to return not only bottles for re-use, but also other items, such as paper, textiles and aluminium drinks cans, as seen here at a Southwark Council site in south London.

A transparent model of an award-winning device that uses hydrocyclonic forces to separate oil from water is examined by one of the engineers who helped develop it. It can typically remove more than 99 per cent of the oil from oily water, the oil showing up as a dark band within the water stream on the right.

Salmon have returned to the Thames in recent decades, following major improvements in river quality. This one, weighing 5 kg, was caught in a trap used by the National Rivers Authority to monitor salmon before being released to continue its journey upstream.

in from 1993, although initially for a limited number of objectives so as to test the system. The paper also proposed a new classification system, against which individual WQOs would be set. It also proposed a revised scheme for the regular overall general assessment of river quality. The proposed scheme consists of a set of use classes, in addition to the requirements of various EC directives. Each WQO would specify, for an individual stretch of river, the use class standards which would apply and the date by which these should be achieved. The NRA would then set discharge consents at levels necessary to achieve the specified quality objective by the set date.

Cleaning Britain's Rivers

Over the past 30 years, notable progress has been made in cleaning up the previously heavily polluted major estuaries of the east coast of England and Scotland—the Thames, Humber, Tees, Tyne and Forth—which now support varied populations of fish and other wildlife. A 25-year scheme supported by the Government and the EC aims to reduce river pollution and improve water quality throughout the Mersey river basin and estuary. Other major schemes in progress include programmes to improve water quality in the Clyde in Scotland and the Lagan in Northern Ireland.

Aquatic life is one of the indicators of water quality, since many species will not inhabit waters that are heavily polluted. There have been some notable achievements in improving the level of aquatic life in major rivers in Britain, most notably in the River Thames. In 1957 a survey indicated that, with the possible exception of eels, there were no fish at all in the 48 km (30 miles) of the Thames in London, which had once been a good fishing river. In 1974, however, an angler hooked the first salmon to be caught for 150 years, and in 1992 over 200 salmon returned to the Thames to spawn. In all, over 110 species of fish have now been found in the tideway. The Tyne, also once heavily polluted, is now one of the foremost salmon rivers in England. Salmon returned to the Clyde in 1983 after disappearing in the 1860s as a result of pollution. The Tees, almost devoid of fish in the 1950s and 1960s, now also supports a resident salmon population.

River Quality Survey

The quality of Britain's rivers, canals and estuaries is measured in the River Quality Survey, which is conducted at five-year intervals. The most recent survey, conducted in 1990, found that some 89 per cent of river and canal lengths in England and Wales were of good or fair quality (64 per cent being good and 25 per cent being fair). There had been a slight net deterioration since 1980 and 1985, in which years 90 per cent of lengths had been good or fair. This apparent drop is largely accounted for by low water levels caused by several preceding years of low rainfall and followed on from a trend of small but steady improvement in river quality between 1958 and 1980. It should also be noted that the slight drop in quality reflected in the 1990 survey was a net drop; some 4,600 km (2,900 miles) of rivers and canals were actually assigned a higher class in 1990 than in 1985. River quality is generally better in Scotland and Northern Ireland than in England and Wales; some 99 per cent of rivers in Scotland and 95 per cent in Northern Ireland were good or fair quality.

Farm Waste

Farming can be a source of pollution. It is important that farm waste is properly dealt with, as it can significantly pollute Britain's freshwaters: undiluted farm slurry can be up to 100 times, and silage effluent up to 200 times, more polluting than raw sewage.

To deal with the problem, new regulations came into force in 1991 setting minimum standards for new farm waste handling facilities. The regulations also empower the NRA to require farmers to improve installations where there is a significant risk of pollution. To help farmers meet the cost of investing in waste handling facilities, the Government makes grant aid available; some £91 million had been paid out by the end of October 1992. It is estimated that, in total, British farmers have spent some £180 million on pollution control equipment.

In addition, the Ministry of Agriculture, Fisheries and Food, together with the Welsh Office and the NRA, makes great efforts to increase farmers' awareness of pollution problems. For example, the Ministry has published codes of good agricultural practice for the protection of water and air and will do so shortly on soil. These give guidance to farmers on, among other things, the planning of the disposal of their farm wastes. Free initial advice is also offered to farmers on pollution issues through the Agricultural Development and Advisory Service (ADAS), a joint agency of the Ministry and of the Welsh Office. Similar efforts are made in Scotland.

An example of other ways in which the Government is helping farmers to combat pollution is a pilot project in four areas—Cheshire, Devon, Dyfed and Somerset—under which individual farmers are given free advice by ADAS in drawing up their own farm waste disposal plans. These plans specify when manure, slurry and other organic wastes can best be applied to land so as to minimise the risk of pollution and get the most from the nutrients contained in the wastes. It is hoped that about 400 farmers will participate in this study.

These measures are showing some encouraging signs of success. The number and severity of farm pollution incidents can be affected by such factors as the weather, with a reduction in the risk of pollution when the weather is unusually dry. Even allowing for this, however, there appears to have been a significant reduction in the number of serious pollution incidents: in 1991 there were 99 such incidents caused by farm waste in England and Wales, compared with 239 the previous year.

Nitrate Pollution

Nitrate can pollute inland waters by leaching from farmland, by water discharge from sewage treatment works and from other miscellaneous activities. Much of this comes from nitrogen already present in the soil, but some also comes from organic and inorganic fertiliser usage. It is recognised that some farming activities, such as ploughing grasslands, can increase the level of nitrate release. In recognition of the concern about human health, the European Community has set a maximum admissible level of nitrate in drinking water. In some parts of Britain nitrate levels in drinking water are close to or above this limit, and the Government is therefore examining the practical impact of controls of nitrate leaching in agriculture. Ten nitrate-sensitive areas (NSAs), covering about 10,700 hectares (27,000 acres) of agricultural land, have been designated. Farmers in these areas can qualify for government payments by changing their farming practices, for example by growing 'green cover' crops in the autumn to take up residual nitrogen or, in some cases, by changing from arable to low-intensity grassland cultivation. The information from the NSAs will provide an important basis in determining the action plans that Britain is required to draw up by December 1995, under an EC directive. These action plans will be implemented within nitrate-vulnerable zones, which are to be designated in 1993.

Sewage Pollution

Some 96 per cent of households in Britain are connected to sewerage networks, the highest proportion in Europe. Untreated sewage contains organic matter and organisms which can be harmful to human health, and can also contain other contaminants, such as heavy metals, if industrial premises discharge their wastes direct to the sewer network. The treatment of sewage at an inland works is normally to secondary standards, which use natural biological processes of purification.

The Government announced in March 1990 that, in future, all significant discharges of sewage into estuaries or coastal waters would receive at least primary treatment, which involves passing the effluent through settlement tanks where much of the organic matter is removed as sewage sludge. This commitment was subsequently embodied in an EC directive on urban waste water treatment. The main objective of the directive is to ensure that all significant discharges of sewage are treated before they are discharged either to inland surface waters, estuaries or coastal waters. Levels of treatment would vary according to the sensitivity of the receiving waters. The directive includes a provision to ban the dumping of sewage sludge at sea from 1998, a commitment that Britain had already made (see p. 48). The dates for compliance with the directive range from 1998 to 2005, depending upon the size and location of the discharge. Separate provisions also refer to discharges from a number of sectors of industry which have a similar nature to domestic sewage.

Sewage Works Compliance

A £1,200 million investment programme by the water industry was announced in 1988 to bring substandard sewage treatment works into compliance with legal limits. In 1986, some 23 per cent of sewage works in England and Wales had not met their required long-term performance measures in their discharge consents; such plants were given time-limited consents in connection with the water industry's investment programme. This number fell to 17 per cent in 1988, and 12 per cent in 1990.

Bathing Waters

The quality of bathing waters is measured by annual testing to establish whether the mandatory coliform bacteria standards of the EC bathing water directive are being met. Each bathing water is sampled at least 20 times, and, in order to comply, only one sample is allowed to exceed the specified level of coliforms. In the 1992 tests of bathing water quality, it was found that 79 per cent of identified bathing waters (358 out of 455) in Britain met the directive's coliform limit, compared with 66 per cent of the then identified bathing waters in 1988. Of more than 18,000 individual samples taken in the 1992 season, 96 per cent met the standard. The Government expects all but a handful of bathing waters to meet the directive's standards by 1995 and to achieve full compliance soon thereafter.

The Government is undertaking a large study into the health risks of sea bathing. Preliminary results suggest that there may be a slight increase in the reported incidence of minor illnesses among people entering the water, although firm conclusions cannot yet be drawn on the effect of bathing water quality on human health.

Marine Environment

Good progress is being made in reducing the amount of toxic substances released into the sea. The Government's target is a reduction of 50 per cent by 1995, and of 70 per cent for the most dangerous substances such as cadmium, mercury, dioxins and lead. Substantial reductions have already been achieved. For example, the discharge of lead into the North Sea fell by 89 per cent between 1985 and 1990, the discharge of cadmium fell by 61 per cent and mercury discharges fell by 51 per cent. Inputs of other hazardous substances also fell: the input of lindane to the North Sea, for example, fell by 30 per cent over the period 1985–90.

Data produced to monitor compliance with environmental quality standards show that in 1979 the average concentration of mercury in cod caught in Liverpool Bay was 0.26 milligrammes per kilogramme; by 1989 this had fallen to 0.15 mg/kg.

Table 5: Metal Content of Material Licensed for Deposition at Sea 1986–90 *tonnes*

	1986	1990
In dredged material		
Zinc	6,090	3,942
Lead	2,587	1,546
Copper	1,364	896
In solid industrial waste[a]		
Zinc	490	441
Lead	249	223
Copper	195	182
In sewage sludge		
Zinc	542	288
Lead	196	129
Copper	171	147

Source: Ministry of Agriculture, Fisheries and Food

[a] These metals are generally bound up in the mineral content of what is largely mining waste, and are therefore not readily available to be taken up by marine life

North Sea Conferences

Britain is a leading participant in the series of North Sea Conferences. This international forum of the countries bordering the North Sea provides the prime focus for the development of Britain's policies on the marine environment. Measures agreed are being applied by Britain to all its coastal waters, not just the North Sea.

The Third North Sea Conference, held in The Hague in the Netherlands in March 1990, agreed a series of measures to protect the marine environment, including:

— the termination of sewage sludge dumping at sea by the end of 1998;

— stopping the incineration of waste at sea by the end of 1990;

— a strengthened approach to nutrient inputs;

— a harmonised approach to the control of dangerous substances;

— the phasing out and destruction of polychlorinated biphenyls (PCBs) by 1999 at the latest; and

— action to protect and conserve wildlife, especially dolphins and porpoises.

Britain was the first country to produce a detailed implementation plan for its commitments under the Third North Sea Conference; this was published in July 1990. The next full meeting of the North Sea Conference is scheduled for 1995, with an interim meeting on certain subjects in 1993.

Pollution from Ships and the Offshore Industry

Prevention of marine pollution from ships is based largely on international conventions drawn up under the auspices of the International Maritime Organization, a UN agency with its headquarters in London. The main such agreement regulating pollution from ships is the International Convention on the Prevention of Marine Pollution from Ships 1973 (known as Marpol), together with its 1978 Protocol; this will also cover such matters as the carriage of packaged hazardous cargoes, although the relevant section has not yet received sufficient ratifications to come into force.

The requirements of international agreements such as Marpol are implemented for British ships by domestic legislation; this is binding not only on all ships in British waters, but also British ships all over the world. Under such legislation, ships have to be fitted with specific pollution control equipment. It makes it an offence for ships of any nationality to discharge oil, oily mixtures or ships' garbage into British territorial waters. It is also an offence for British-registered ships to make similar discharges into the sea anywhere in the world, except in accordance with the regulations. Enforcement of these regulations is undertaken by the Department of Transport.

As an example of the effect of such international action, Britain introduced new restrictions on dumping ships' rubbish overboard close to the coastline in December 1988; this was backed up by a publicity campaign to alert ships' crews to the new regulations.

Offshore oil operators must ensure that oil does not escape into the sea and are required to have contingency plans for dealing with oil spilled accidentally. Discharges containing oil are controlled under the Prevention of Oil Pollution Act 1971. The Department of Transport's Marine Pollution Control Unit (see p. 52) runs a programme, funded by the Department of Trade and Industry, of overflights of British oil and gas rigs by its pollution-detection aircraft. The North Sea Conference has agreed to eliminate progressively pollution caused by discharges of oil-contaminated drilling cuttings from offshore platforms, by 1994 in the case of single wells and in due course for all wells. Discharges of oil in drill cuttings from Britain's oil industry fell from 18,500 tonnes in 1988 to 12,320 tonnes in 1990.

Sea Disposal

The environment of the north-east Atlantic is currently protected by the Oslo and Paris Conventions. The former was signed in 1972 and came into force in 1974; it sets guidelines designed to control marine pollution caused by dumping from ships and aircraft and also deals with the incineration of waste at sea. The present Paris Convention was signed in 1974 and came into force in 1978; it deals with discharges from land or offshore platforms directly into the sea, discharges from rivers, and discharges into the air which can eventually fall into the sea. The conventions work through two commissions which meet annually. In 1990 it was decided to review the conventions at ministerial level by 1992 and produce a revised convention; a new convention was agreed in Paris in September 1992. This agreement:

—establishes an effective framework for future protection of the marine environment from pollution;

—commits all the countries to the use of best available techniques and best environmental practice to protect the marine environment;

—establishes a ban on all forms of dumping at sea, with very few listed exceptions;

—recognises that inputs of hazardous substances, in particular organohalogen substances, should be reduced to levels not harmful to mankind or nature by the year 2000; and

—gives high priority to monitoring and periodic assessment of the quality of the marine environment of the north-east Atlantic as a basis for future decisions.

A licence has to be obtained for the deposit of any substance or article into the sea and tidal waters below the high water mark. Disposal at sea is presently permitted where no harm to the marine environment can be shown and where there are no practicable alternatives on land. Nevertheless, as a precautionary measure, a timetable for ending the disposal of industrial waste at sea was announced by the Government in 1990. The Government's intention is that no industrial waste should be dumped at sea beyond early 1993. In accordance with the agreement at the Third North Sea Conference, waste has not been licensed for incineration at sea since 1990. Tougher powers to control waste dumping at sea were included in the Environmental Protection Act 1990, closing a loop-hole in earlier legislation.

Sewage Sludge

The Government is committed to stopping the sea dumping of sewage sludge (the residue left after treatment) by the end of 1998; this decision was announced in March 1990 and subsequently embodied in an EC directive. Although some 70 per cent of sewage sludge is already disposed of onshore, this will require the development of alternative uses and disposal methods for sewage sludge. These alternatives could include greater use of sewage sludge in agriculture and more incineration, but there are difficulties to be overcome—for example, normal planning procedures have to be followed to obtain permission for the construction of incinerators.

Radioactive Waste

Britain stopped the sea disposal of low-level radioactive waste in 1982, and in 1988 announced that the disposal of drums of radioactive waste would not be resumed. The Government did not rule out the option of the future disposal of large items such as boilers from decommissioned nuclear power stations, but undertook to keep under review the necessity of retaining this option for the future. Such disposal is only a possibility, but the option cannot not be closed off at present, however, because there is not yet a workable land-based method of disposal.

However, Britain has agreed in the new convention drawn up in September 1992 (see p. 50) to a 15-year ban on dumping all radioactive waste in the area covered by the convention. After that period Britain will be free to exercise the option to dispose of bulky low- and intermediate-level items. The conditions attached mean that this will not be exercised without an exhaustive discussion with all other contracting parties. This would require full analyses of hazards to human health, harm to living resources and marine ecosystems, damage to amenities and interference with other legitimate uses of the sea. The only radioactive wastes for which Britain could envisage the possibility of sea disposal in future are large bulky items arising from decommissioning of nuclear facilities, for which such disposal might prove to be the best environmental option when account was taken of the radiation that would be received by people involved in cutting up such items for land deposit.

Marine Pollution Control Unit

The Department of Transport's Marine Pollution Control Unit (MPCU) is responsible for dealing with pollution at sea when oil or other dangerous substances from a shipping casualty threaten major coastal pollution. It maintains a national contingency plan and dispersant spraying, cargo transfer, mechanical recovery and beach-cleaning resources. It also has two remote sensing aircraft capable of detecting possible illegal discharges and of quantifying oil pollution. In 1991 the MPCU received 220 reports of pollution, of which 38 were detected by its aircraft. None of these was major and only 23 required further action by the MPCU. The Ministry of Agriculture, Fisheries and Food authorises all new oil dispersants for use in British waters, and regularly reviews all previously authorised ones.

The MPCU's plans were put to the test in January 1993 as a result of the grounding of the MV *Braer* in heavy weather off Sumburgh in the Shetland Isles. Its cargo of 80,000 tonnes of oil was spilled, making it the largest spillage of oil in British waters since the loss of the *Torrey Canyon* in 1967. The MPCU deployed both dispersant spraying and remote sensing aircraft to the area, although in the event gale force winds largely dispersed the oil and little spraying was necessary.

Land Pollution and Waste Management

Strict controls are applied to protect the environment from such problems as litter, and incorrect disposal of waste. As in other areas, the Environmental Protection Act 1990 has significantly improved standards.

Waste Collection and Disposal

The disposal of controlled waste on land is governed by the Control of Pollution Act 1974 and the Environmental Protection Act 1990. Controlled waste, however, excludes waste from agricultural premises, mining and quarrying waste and (in certain circumstances) sewage sludge, while radioactive waste is controlled under other provisions. The exclusion of agricultural, mining and quarrying waste is currently under review. Under the legislation, local authorities have functions as waste collection authorities, waste disposal authorities and waste regulation authorities.

Waste collection authorities, which in England and Wales are the district councils, have a duty to collect household waste and to offer a commercial waste collection service (although many businesses instead enter a refuse collection arrangement with a private contractor). Waste disposal authorities (in non-metropolitan England, the county councils; in Wales the district councils) are responsible for ensuring the proper disposal of the refuse collected. In Scotland, the district and islands councils act as both waste collection authorities and waste disposal authorities. In Northern

Ireland, responsibility for the collection, disposal and regulation of
waste rests with the district councils.

Table 6: Waste Reception and Disposal Facilities in Britain

Type of facility	Number of disposal licences
Landfill	4,196
Civic amenity	559
Transfer[a]	936
Storage[b]	274
Treatment[c]	122
Incineration	212
Other	366

Sources: Department of the Environment, the Scottish Office Environment
Department, Department of the Environment for Northern Ireland.

[a]Facilities licensed for receipt, sorting, consolidation and onward movement of
waste.
[b]Facilities licensed principally for storing waste remotely from final disposal.
[c]Including physical, chemical and biological treatment and solidification.

Waste regulation authorities are responsible for regulating the
disposal of controlled wastes. They are required to draw up and
revise periodically a waste disposal plan. Legislation has also established a licensing system for waste disposal sites, treatment plants
and storage facilities receiving controlled wastes, while a more
intensive control system applies for certain especially difficult
wastes. Operators of landfill sites are required to obtain a licence
from the relevant waste regulation authority, which would contain
conditions designed to ensure that there is no harm to public health
or water resources and no serious detriment to local amenities.
Under the Environmental Protection Act 1990, these safeguards
are being strengthened (see p. 57). The Hazardous Waste
Inspectorate for Scotland may advise local authorities on how to

improve their control of waste management and on how to work towards environmentally acceptable standards for dealing with hazardous wastes. In Northern Ireland, similar advice is offered to district councils by the Environment Service.

Landfill

Landfill has not generally caused major problems as a means of waste disposal, partly as a result of suitable licence conditions being imposed upon operators to guard against pollution. It is important, however, that proper controls are maintained and suitable precautions taken. For example, liquid leaching from landfill sites could contaminate nearby groundwaters, so licences would normally stipulate the use of liners or bunds to prevent this. Landfill may also give rise to gases, principally methane arising from the decomposition of organic matter. If not properly dealt with, this could present a hazard. Sometimes methane is dealt with by being vented or flared off, but increasingly there is interest in recovering it as a potentially valuable source of energy.

Incineration

Owing to increased costs anticipated for landfill (due partly to shortages of suitable sites and partly to the cost of meeting higher environmental standards), there has been a move towards incineration as a disposal option. Most new incinerators are designed to use the heat from burning rubbish to generate electricity. Such installations would require authorisation under IPC to ensure that the best available techniques not entailing excessive cost are used to prevent or minimise potentially harmful releases. Incineration sterilises the material for final disposal and reduces it in volume by about 90 per cent. Some 3 million tonnes of municipal waste is incinerated each year.

Improved Controls
Part II of the Environmental Protection Act 1990, which is being phased in from April 1992 onwards, strengthens existing curbs on waste disposal. Responsibility for proper handling of waste has been imposed on everyone who has control of it from production to final disposal or reclamation. Authorities will be able to refuse licences if the applicant is not a fit and proper person. Operators will be responsible for their sites until the waste regulation authority is satisfied that there is no likelihood of future risk to human health or the environment. In England and Wales local authorities' waste disposal operations are to be transferred to 'arm's length' companies or private contractors, so as to separate them from the authorities' other jobs of setting policies and standards, and enforcement. Subject to further legislation, the Government intends that its proposed Environment Agency would take over local authority waste regulation functions.

Litter
It is a criminal offence to leave litter in any public place in the open air or to dump rubbish except in designated places. The maximum penalty for this, previously set at £400, was increased under the Environmental Protection Act 1990 to £1,000 (it has since been raised again to £2,500). The Act also introduced new powers for the issue of litter abatement orders and new duties on local authorities to keep their public land free of litter and refuse (including dog faeces). Similar powers are expected to come into force in Northern Ireland during 1993. The duty to keep public land clear of litter also extends to certain other areas, such as railway embankments and cuttings.

To help counteract the problem of litter, the Government gives financial support to the Tidy Britain Group (£2.8 million in

1992–93), which provides a comprehensive litter abatement programme in collaboration with local authorities. The Group secures sponsorship from industry to undertake litter abatement promotions and programmes such as its Neighbourhood Litter Watch scheme. The Group's activities were extended to Northern Ireland with the establishment in 1990 of Tidy Northern Ireland.

A litter survey carried out by the Tidy Britain Group in March 1992—exactly one year after the implementation of the relevant sections of the 1990 Act—showed that Britain's streets were about 13 per cent cleaner in 1992 than in the previous year.

Recycling and Materials Reclamation

Recycling materials from household waste can have considerable benefits for the environment. Not only are resources used more efficiently, but the need for landfill space is reduced. The Government encourages the reclamation and recycling of waste materials wherever this makes environmental sense; its target is for half of all recyclable household waste to be recycled by the year 2000. In other areas, more materials have traditionally been reclaimed: it is estimated, for example, that some 70 to 75 per cent of the material in scrap cars (mostly metals) is recovered.

Under the Environmental Protection Act 1990, local authorities are required to make plans for the recycling of waste. By January 1992, over 340 such plans had been received from local authorities in England, out of a total of 366 authorities. In Wales, a third of the plans had been received by October 1992.

The Government has supported pilot 'Recycling City' initiatives in Sheffield, Cardiff, Dundee and the county of Devon, which have tested a variety of collection and sorting methods. In these and other locations, trial collections of recyclable waste are being

supported and monitored. Members of the public can deposit used
glass containers for recycling in bottle banks—there are over 7,500
such sites in Britain, and it is anticipated that there will be 10,000
by 1995. In addition, there are similar can banks (of which there are
over 700), paper banks and, in some cases, plastics or textiles banks
to collect these materials for recycling. There are at least 33 kerb-
side collection schemes, serving 700,000 households at least once a
month. In addition, voluntary organisations also arrange collections
of waste material. They may be paid recycling credits by the relevant waste disposal authority for doing so.

Discussions with industry aimed at improving markets for
collected recyclable waste have led to voluntary targets, such as a 40
per cent recycled content of newspapers by the year 2000, and grant
aid for research and new plant. Encouraging research into recycling
technologies is one of the priorities of the Government's DEMOS
scheme (see p. 20). Other initiatives have also been set up, for
example:

—a joint government–industry working party, announced in July
   1992, to promote the recycling of scrap cars in Britain, which is
   important given the rise in plastics in car components;

—a process developed in Britain for recycling old car tyres into oil
   and carbon, by decomposing them using pyrolysis; and

—a lorry organised by one city council to collect materials that
   need to be treated specially for re-use, such as old tins of
   paint.

The Government supports local authority recycling initiatives, for example by means of supplementary credit approvals; in
England nearly £15 million was made available by the Department
of the Environment in this way in 1992–93.

Recycling Credits

Financial incentives for recycling were provided with the introduction of recycling credits in the Environmental Protection Act 1990; the relevant section came into force in April 1992. Under this scheme, waste disposal authorities in England and Wales have a duty to pay recycling credits to waste collection authorities to reflect the savings made in disposal costs because of waste being removed for recycling; they also have the power to pay recycling credits to other recycling collectors. It was announced in September 1992 that the level of these payments would double in April 1994, thus increasing the financial inducement to separate waste for recycling; whereas the credits currently reflect half the saving of waste disposal costs made by recycling, from this date the full saving will have to be passed on to the waste collection authority. In Scotland, waste disposal authorities and waste collection authorities are not separate. However, waste disposal authorities may voluntarily pay recycling credits to third parties.

Hazardous Waste

A relatively small proportion of waste is classified as special waste—about 2 per cent of total controlled waste arisings. However, because some substances, such as PCBs, dioxins and heavy metals, pose very significant hazards to human health, wastes which could contain such materials require very careful treatment.

The best way to treat hazardous waste will vary according to its exact nature. For some substances, such as pesticides residue and PCBs, high-temperature incineration is the most suitable method of disposal, and there are some specialised incinerators in Britain capable of this. For other less hazardous industrial wastes, co-disposal may be suitable—this is disposal in landfills also taking

household and similar waste. Natural biological and chemical processes within the waste mass can be employed to change the nature
of the waste and render it benign.

Some hazardous waste is imported into Britain to be dealt
with. By law such international shipments have to be notified to the
competent authorities in each country and accompanied by full
documentation with such details as the volume and nature of the
waste, its origins and final destinations. Although these regulations
derive from EC directives, they apply to all international movements of hazardous waste, not just those starting in other member
states. The Government believes that developed countries should
become self-sufficient in disposing of their own wastes, but that it
is preferable for hazardous waste arising in developing countries to
be sent to countries with suitable facilities to be dealt with properly. Developing countries are unlikely to possess the necessary
facilities to do this at home.

Hazardous Substances

When a new chemical is placed on the market in the EC for the first
time, the manufacturer or importer must provide detailed information on its properties. In Britain, the Health and Safety Executive
(HSE) and the Department of the Environment then evaluate its
potential to harm mankind or the environment, and share this
information with other members of the Community. If a chemical
is found to be potentially hazardous, appropriate action is taken to
control it.

Some 110,000 different chemicals were marketed in the EC
before 1981. About 5,000 of these are produced in quantities over
10 tonnes a year, and about 1,500 in quantities over 1,000 tonnes a
year. There is little information about the environmental effects of

many industrial chemicals which are produced in large volumes, but Britain is taking steps to evaluate and, if necessary, control them. The European Commission has proposed legislation which would require all chemicals already on the market in large quantities to be thoroughly evaluated and, if necessary, tested. A priority list of chemicals on which more information is needed would be drawn up. The manufacturers or importers of these priority chemicals would be required to supply the information necessary to evaluate the risk to human health and the environment, and carry out any necessary testing.

A government report published in January 1992 shows that the level of dioxins in British food is so low that they can only be detected by the most sophisticated techniques capable of measuring accurately to ten parts in one thousand million million.

Pesticides

Pesticides are strictly controlled in Britain; from April 1993 the Pesticides Safety Division of the Ministry of Agriculture, Fisheries and Food became an executive agency: the Pesticides Safety Directorate. The Government is assisted in the regulation and approval of pesticides by an independent Advisory Committee on Pesticides, established as a statutory body under the Food and Environment Protection Act 1985. There is a large and systematic programme under way to review older pesticides which were approved some time ago. It is intended to implement an EC collaborative programme jointly to review the use of older pesticides in the near future.

Through its Working Party on Pesticide Residues, the Government undertakes a considerable programme of monitoring of pesticide residue levels in food to ensure that these are not a danger to human health. The results for the 1991 programme, in which over 2,500 samples of the foodstuffs most likely to contain residues were analysed, showed that 71 per cent of samples contained no detectable pesticide residue, while in only 1 per cent of cases were levels above the set maximum residue levels. These levels are based on good agricultural practice and take account of consumer safety. They are not themselves safety levels and it does not follow that food is unsafe if the levels are exceeded. A Total Diet Study carried out in 1989–90 showed a continued decline in residues of persistent organochlorine pesticide residues in the diet, and demonstrated that estimated average dietary intakes were well within acceptable levels set by the United Nations Food and Agriculture Organisation and the World Health Organisation.

Government research also looks at ways of minimising pesticide usage. For example, the Ministry of Agriculture, Fisheries and Food's Boxworth Project has paved the way for new arable farming systems which reduce the use of pesticides, having shown that:

—money spent on preventative use of pesticide is often not recovered by extra crop yields;

—many beneficial insects that eat pests are also very vulnerable to high inputs of pesticides; and

—leaving margins around the field can make an important 'reservoir' for such useful predators.

Further studies are building on this work. Other schemes have been set up; for example, the agrochemical sector and the supply industry ran a national pesticides retrieval scheme, under which farmers could have their old and unwanted pesticides removed cheaply for proper disposal. Government guidelines on minimising the non-agricultural use of herbicides were published in December 1992.

Britain is also working with developing countries to strengthen their capacity to deal with concerns over the export, import and use of hazardous chemicals, especially pesticides.

Radioactivity

Man-made radiation represents only a small fraction of that to which the population is exposed; most is naturally occurring. Nevertheless, that fraction is subject to stringent control. Users of radioactive materials must be registered by HMIP in England and Wales, and equivalents in Scotland and Northern Ireland, and authorisation is also required for the accumulation and disposal of radioactive waste. The main legislation is the Radioactive Substances Act 1960, as amended by the Environmental Protection Act 1990. The HSE, through its Nuclear Installations Inspectorate, is the authority responsible for the granting of nuclear site licences

Table 7: Radiation Exposure of the British Population

	%
Natural	87
of which	
Radon	51
Gamma rays	14
Internal	12
Cosmic rays	10
Artificial	13
of which	
Medical	12
Other	1

Source: National Radiological Protection Board.

for major nuclear installations. No installation may be constructed or operated without a licence granted by the Executive.

The National Radiological Protection Board (NRPB) provides an authoritative point of reference on radiological protection. The Department of the Environment's 1992–93 budget for research into environmental radioactivity and radioactive waste is £5.6 million, and that for the Ministry of Agriculture, Fisheries and Food is £2.8 million.

Radioactive Waste Disposal

Radioactive wastes vary widely in nature and level of activity, and the methods of disposal reflect this. Some very low-level radioactive wastes can be disposed of safely in the same way as other industrial and household wastes. UK Nirex Ltd—a partnership in which the shares are owned by the Government, British Nuclear Fuels plc, Nuclear Electric plc and Scottish Nuclear Ltd—[6]is responsible for carrying out the disposal of low-level and intermediate-level radioactive wastes. Decisions on Britain's strategy for radioactive waste disposal remain with the Government.

At present, most low-level waste is disposed of at a facility at Drigg in Cumbria. Nirex is, however, developing plans for a deep disposal facility for solid low-level and intermediate-level radioactive waste. The preferred design for this consists of an access tunnel spiralling downwards to disposal chambers half a mile underground, together with a waste handling building on the surface. In July 1991 Nirex announced that its site investigations for

[6]Nuclear Electric plc and Scottish Nuclear Ltd are the companies that were set up in preparation for the privatisation of the electricity supply industry; they own and operate Britain's nuclear power stations formerly belonging to the Central Electricity Generating Board. They remain in state ownership. For further details see *Energy and Natural Resources* (Aspects of Britain: HMSO, 1992).

this were being concentrated on an area adjacent to the British Nuclear Fuels site at Sellafield (Cumbria). As part of this work Nirex intends to construct an underground rock characterisation facility, which would supplement information gained on an earlier borehole drilling programme. The current target date for commissioning the deep disposal facility is 2007, although that could be held up by a variety of factors, such as delays in obtaining the necessary planning consents.

The Department of the Environment is sponsoring research, in collaboration with other countries, into disposal of high-level or heat-generating waste. This waste will first be stored, preferably in vitrified form, for at least 50 years to allow the heat and radioactivity to decay. A vitrification plant has been opened at the Sellafield site to provide this service.

In July 1992, following public consultation, the NRPB published principles that should be observed in the land-based disposal of solid radioactive wastes. The guidance, which updated previous advice issued in 1983, stipulated that:

— the radiological risk from a particular disposal facility should not exceed one chance in 100,000 per year of serious health effects to the most exposed individuals or their descendants;

— future generations should receive protection at least equivalent to that afforded to members of the public alive at present; and

— the risk to the public should be as low as reasonably achievable.

The NRPB has offered guidance on implementing these principles.

Radioactive Discharges
Radioactive discharges require authorisation from both HMIP and the Ministry of Agriculture, Fisheries and Food in England, and

from the Welsh and Scottish Offices respectively for installations in Wales and Scotland. These consents are subject to very stringent limits set in the authorisation programme, designed to protect the health of the public by limiting the exposure of defined 'critical populations'—the group of people most at risk from discharges from a particular nuclear facility.

British Nuclear Fuels has undertaken substantial investment to reduce discharges from its Sellafield site; over the past few years waste management and effluent treatment facilities costing £2,000 million were completed there. As a result, discharges of radioactivity to the sea from Sellafield are about 1 per cent of their level in the 1970s. Further long-term reductions in discharges are expected after the new Enhanced Actinide Removal Plant starts operating in 1993. This will remove the most radiologically significant nuclides, especially americium and plutonium.

Radon

Radon is a naturally occurring radioactive gas deriving from the radioactive decay of uranium. It occurs to some extent in all soils, but is particularly prevalent in and around areas where igneous rocks, such as granite, are found. In the open it disperses rapidly, but it can pose a hazard to human health where it seeps into houses and accumulates, a danger that well-insulated modern homes are particularly prone to. It is thought that one in twenty lung cancers in Britain may be due to radon.

In 1987 the Government announced measures to deal with the problem, including a free survey by the NRPB for householders living in radon-affected areas. The results are confidential, and can indicate whether or not remedial action is advisable. Such remedial action can include measures to ventilate the house better or to stop radon seeping in from the soil. In 1990, following new advice from the NRPB as to the level at which radon could pose a serious health concern, the Government halved the 'action level'— the concentration at which it is recommended that householders take action to reduce radon in their homes. This action level is now set at 200 becquerels[7] per square metre. The NRPB initially designated Cornwall and Devon as a 'radon affected area' (an area where more than 1 per cent of the houses exceed the action level); over 80,000 requests for surveys have been received in this area. In July 1992, Northamptonshire and parts of Derbyshire and Somerset were also designated as radon affected areas. A publicity campaign covering Northamptonshire and much of Derbyshire was launched in October 1992 to alert householders to the problem and encourage them to take advantage of the free survey. The NRPB has also carried out some surveys of radon levels in Welsh homes. The Government has published a booklet, *The Householders' Guide to Radon*, an updated version of which came out in October 1992.

Monitoring

There is a large government programme to monitor radioactivity and the potential exposure of members of the public to it. In England and Wales, the Ministry of Agriculture, Fisheries and Food is largely responsible for this.

Radioactive materials can be particularly harmful to health when ingested. To complement its other monitoring programmes, the Ministry therefore carries out a monitoring programme to ensure that radioactivity does not enter the food chain in levels harmful to human health. The 1991 results, which derived from

[7]The becquerel is the internationally-recognised unit of radioactivity; one becquerel is defined as one radioactive decay per second.

over 6,000 samples of milk, fruit, vegetables and meat, showed that the exposure of consumers to radioactivity via the food chain remained well below internationally recommended limits. The maximum potential exposure via the food chain was about 6 per cent of the annual limit.

Following the accident at the Chernobyl nuclear power station in the then Soviet Union in 1986, the Government set up a national radiation monitoring network and overseas nuclear accident response system, known as RIMNET. An interim version has been operating since 1988. This is scheduled to be replaced during 1993 by a larger system with 92 monitoring stations across Britain. The Ministry of Agriculture, Fisheries and Food also continues its monitoring of marked sheep leaving the areas restricted after Chernobyl, and undertakes regular scientific surveys in these areas.

The Ministry of Agriculture, Fisheries and Food's programme for the terrestrial environment is complemented by one for the aquatic environment. Samples of fish, shellfish, sea water, sediments and seaweed are collected and analysed for their radioactive content. In addition, direct radiation is measured in intertidal areas using portable instruments. The maximum potential exposure to man-made radioactivity from aquatic pathways in 1991 was about one-sixth of the internationally recommended limit.

Genetically Modified Organisms

The Environmental Protection Act 1990 introduced stronger controls over genetically modified organisms (GMOs). Anyone who intends to carry out activities with GMOs, other than approved GMO products, first has to do an environmental risk assessment and, in some cases, first notify the Government or obtain its consent. The Government has powers to prevent activities with

GMOs where this would involve a significant risk of damage to the environment, and inspectors have the power to destroy or render harmless organisms which are likely to cause damage to the environment. A unified system of administrative control for GMOs is being developed in parallel with these provisions, involving the Department of the Environment, the HSE and other interested departments. Britain has taken the lead in developing guidelines on biotechnology safety, which were approved at the UNCED meeting in June 1992. Britain will continue to press for an international convention on risk assessment and management for modern biotechnology.

Noise

Local authorities have a duty to inspect their areas for noise nuisance and to take reasonably practical steps to investigate complaints. They must serve a noise abatement notice on people creating a noise nuisance. They can also designate 'noise abatement zones' within which registered levels of noise from specified premises may not be increased without their permission. There are specific provisions in law to:

—control noise from construction and demolition sites;

—control the use of loudspeakers in the streets; and

—enable individuals to take independent action through the courts against noise amounting to a nuisance.

Tougher measures against noise were proposed in the environment White Paper, such as the introduction of statutory controls over burglar alarms, improving the requirements for sound insulation in the building regulations, and the strengthening of local authority powers to deal with noise nuisance. Most of its commitments were endorsed by an independent working party set up by the Government to review noise control, which reported in October 1990. Government spending on environmental noise research is expected to be about £800,000 in 1992–93. The Government also provides financial assistance for noise mediation schemes.

Transport is a major source of noise, and control measures are aimed at reducing it at source, through requirements limiting the noise that aircraft and motor vehicles may make, and at protecting people from its effects. Regulations set out the permissible noise levels for various classes of new vehicle. Government research also looks at ways of reducing noise, and has demonstrated the feasibility of tougher noise limits for heavy goods vehicles.

Compensation may be payable for loss in property values caused by physical factors, including noise from new or improved public works such as roads, railways and airports. Regulations also enable highway authorities to carry out or make grants for the insulation of homes that would be subject to specified levels of increased noise caused by new or improved roads. Noise insulation may be provided where construction work for new roads may seriously affect nearby homes.

Britain has played a leading role in negotiations aimed at the gradual phasing out of older, noisier subsonic jet aircraft. Flying non-noise-certificated aircraft has been banned in Britain, and since 1990 British operators have no longer been allowed to add to their fleets further 'Chapter 2' aircraft (noisier planes, as classified by international agreement). A complete ban on the operation of Chapter 2 aircraft will begin to be implemented in April 1995, and it is intended to phase out all these types by April 2002. Various operational restrictions have been introduced to reduce noise disturbance further at Heathrow, Gatwick and Stansted, where the Secretary of State for Transport has assumed responsibility for noise abatement. These measures include:

—restrictions on the type and number of aircraft operating at night;

—the routeing of departing aircraft on noise preferential routes; and

—quieter take-off and landing procedures.

The population affected by aircraft noise at Heathrow fell from nearly 1.5 million in 1978 to less than 550,000 in 1988, even though the number of air transport movements increased by about a quarter. This was largely because of the phasing out of older, noisier aircraft.

The Government is reviewing arrangements for noise mitigation at smaller airfields.

List of Abbreviations

ACBE	Advisory Committee on Business and the Environment
ADAS	Agricultural Development and Advisory Service
AFRC	Agricultural and Food Research Council
CFCs	chlorofluorocarbons
CO	carbon monoxide
CO_2	carbon dioxide
DEMOS	Environmental Management Options Scheme
EC	European Community
ETIS	Environmental Technology Innovation Scheme
GMOs	genetically modified organisms
HCFCs	hydrochlorofluorocarbons
HMIP	Her Majesty's Inspectorate of Pollution
HMIPI	Her Majesty's Industrial Pollution Inspectorate
HSE	Health and Safety Executive
IPC	integrated pollution control
LAAPC	local authority air pollution control
MPCU	Marine Pollution Control Unit
NERC	Natural Environment Research Council
NO_x	oxides of nitrogen
NRA	National Rivers Authority
NRPB	National Radiological Protection Board
NSAs	nitrate-sensitive areas
ODA	Overseas Development Administration
PCBs	polychlorinated biphenyls
SO_2	sulphur dioxide
TIGER	Terrestrial Initiative in Global Environmental Research
UNCED	United Nations Conference on Environment and Development
VOCs	volatile organic compounds
WQOs	water quality objectives

Addresses

Department of the Environment, 2 Marsham Street, London SW1P 3EB.

Department of Trade and Industry, Ashdown House, 123 Victoria Street, London SW1E 6RB.

Department of Transport, 2 Marsham Street, London SW1P 3EB.

Ministry of Agriculture, Fisheries and Food, 3 Whitehall Place, London SW1A 2HH.

Northern Ireland Environment Service, Calvert House, 23 Castle Place, Belfast BT1 1FY.

The Scottish Office, St Andrew's House, Edinburgh EH1 3DE.

Welsh Office, Cathays Park, Cardiff CF1 3NQ.

Her Majesty's Inspectorate of Pollution, Romney House, 43 Marsham Street, London SW1P 3DY.

National Radiological Protection Board, Chilton, Didcot, Oxfordshire OX11 0RQ.

National Rivers Authority, Rivers House, Waterside Drive, Aztec West, Almondsbury, Bristol BS12 4UD.

Natural Environment Research Council, Polaris House, North Star Avenue, Swindon, Wiltshire SN2 1EU.

United Kingdom Nirex Ltd, Curie Avenue, Harwell, Didcot, Oxfordshire OX11 0RH.

Further Reading

			£
Environment in Trust.	Department of the Environment	1991	Free
Environmental Directory. ISBN 1 870257 05 7.	Civic Trust	1988	4.00
Green Rights and Responsibilities in Scotland: a Citizen's Guide to the Environment.	Scottish Office	1992	Free
Householder's Guide to Radon.	Department of the Environment	1992	Free
Pesticides, Cereal Farming and the Environment: the Boxworth Project. ISBN 0 11 242876 2.	HMSO	1992	45.00
The UK Environment. ISBN 0 11 752420 4.	HMSO	1992	14.95
The Welsh Environment: a Guide to Your Rights and Responsibilities.	Welsh Office	1992	Free
This Common Inheritance: Britain's Environmental Strategy. Cm 1200. ISBN 0 10 112002 8.	HMSO	1990	27.00
This Common Inheritance: the First Year Report. Cm 1655. ISBN 1 10 116552 8.	HMSO	1991	21.00
This Common Inheritance: the Second Year Report. Cm 2068. ISBN 0 10 120682 8.	HMSO	1992	21.00

Annual Statistics
Digest of Environmental Protection and Water Statistics.   HMSO

The Scottish Environment Statistics.	Scottish Office
Environmental Digest for Wales.	Welsh Office
Annual Report of the Working Party *on Pesticide Residues.*	HMSO

Index

Printed in the UK for HMSO.
Dd.0296517, 5/93, C30, 51-2423, 5673, 240664.

BRITAIN HANDBOOK

The annual picture of Britain is provided by *Britain: An Official Handbook* - the forty-fourth edition will be published early in 1993. It is the unrivalled reference book about Britain, packed with information and statistics on every facet of British life.

With a circulation of over 20,000 worldwide, it is essential for libraries, educational institutions, business organisations and individuals needing easy access to reliable and up-to-date information, and is supported in this role by its sister publication, *Current Affairs: A Monthly Survey.*

Approx. 500 pages; 24 pages of colour illustrations; 16 maps; diagrams and tables throughout the text; and a statistical section. Price £19·50.

Buyers of Britain 1993: An Official Handbook *have the opportunity of a year's subscription to* Current Affairs *at 25 per cent off the published price of £35·80. They will also have the option of renewing their subscription next year at the same discount. Details in each copy of* Handbook, *from HMSO Publications Centre and at HMSO bookshops (see back of title page).*